1 MONTH OF FREE READING

at
www.ForgottenBooks.com

By purchasing this book you are eligible for one month membership to ForgottenBooks.com, giving you unlimited access to our entire collection of over 1,000,000 titles via our web site and mobile apps.

To claim your free month visit:
www.forgottenbooks.com/free923323

* Offer is valid for 45 days from date of purchase. Terms and conditions apply.

ISBN 978-0-260-02903-4
PIBN 10923323

This book is a reproduction of an important historical work. Forgotten Books uses state-of-the-art technology to digitally reconstruct the work, preserving the original format whilst repairing imperfections present in the aged copy. In rare cases, an imperfection in the original, such as a blemish or missing page, may be replicated in our edition. We do, however, repair the vast majority of imperfections successfully; any imperfections that remain are intentionally left to preserve the state of such historical works.

Forgotten Books is a registered trademark of FB &c Ltd.
Copyright © 2018 FB &c Ltd.
FB &c Ltd, Dalton House, 60 Windsor Avenue, London, SW19 2RR.
Company number 08720141. Registered in England and Wales.

For support please visit www.forgottenbooks.com

A Method for Designing Multi-Screw Waveguide Tuners

NBS TECHNICAL PUBLICATIONS

PERIODICALS

JOURNAL OF RESEARCH reports National Bureau of Standards research and development in physics, mathematics, chemistry, and engineering. Comprehensive scientific papers give complete details of the work, including laboratory data, experimental procedures, and theoretical and mathematical analyses. Illustrated with photographs, drawings, and charts.

Published in three sections, available separately:

● **Physics and Chemistry**

Papers of interest primarily to scientists working in these fields. This section covers a broad range of physical and chemical research, with major emphasis on standards of physical measurement, fundamental constants, and properties of matter. Issued six times a year. Annual subscription: Domestic, $9.50; foreign, $11.75*.

● **Mathematical Sciences**

Studies and compilations designed mainly for the mathematician and theoretical physicist. Topics in mathematical statistics, theory of experiment design, numerical analysis, theoretical physics and chemistry, logical design and programming of computers and computer systems. Short numerical tables. Issued quarterly. Annual subscription: Domestic, $5.00; foreign, $6.25*.

● **Engineering and Instrumentation**

Reporting results of interest chiefly to the engineer and the applied scientist. This section includes many of the new developments in instrumentation resulting from the Bureau's work in physical measurement, data processing, and development of test methods. It will also cover some of the work in acoustics, applied mechanics, building research, and cryogenic engineering. Issued quarterly. Annual subscription: Domestic, $5.00; foreign, $6.25*.

TECHNICAL NEWS BULLETIN

The best single source of information concerning the Bureau's research, developmental, cooperative and publication activities, this monthly publication is designed for the industry-oriented individual whose daily work involves intimate contact with science and technology—*for engineers, chemists, physicists, research managers, product-development managers, and company executives*. Annual subscription: Domestic, $3.00; foreign, $4.00*.

* Difference in price is due to extra cost of foreign mailing.

NONPERIODICALS

Applied Mathematics Series. Mathematical tables, manuals, and studies.

Building Science Series. Research results, test methods, and performance criteria of building materials, components, systems, and structures.

Handbooks. Recommended codes of engineering and industrial practice (including safety codes) developed in cooperation with interested industries, professional organizations, and regulatory bodies.

Special Publications. Proceedings of NBS conferences, bibliographies, annual reports, wall charts, pamphlets, etc.

Monographs. Major contributions to the technical literature on various subjects related to the Bureau's scientific and technical activities.

National Standard Reference Data Series. NSRDS provides quantitive data on the physical and chemical properties of materials, compiled from the world's literature and critically evaluated.

Product Standards. Provide requirements for sizes, types, quality and methods for testing various industrial products. These standards are developed cooperatively with interested Government and industry groups and provide the basis for common understanding of product characteristics for both buyers and sellers. Their use is voluntary.

Technical Notes. This series consists of communications and reports (covering both other agency and NBS-sponsored work) of limited or transitory interest.

Federal Information Processing Standards Publications. This series is the official publication within the Federal Government for information on standards adopted and promulgated under the Public Law 89–306, and Bureau of the Budget Circular A–86 entitled, Standardization of Data Elements and Codes in Data Systems.

CLEARINGHOUSE

The Clearinghouse for Federal Scientific and Technical Information, operated by NBS, supplies unclassified information related to Government-generated science and technology in defense, space, atomic energy, and other national programs. For further information on Clearinghouse services, write:

Clearinghouse
U.S. Department of Commerce
Springfield, Virginia 22151

Order NBS publications from: Superintendent of Documents
Government Printing Office
Washington, D.C. 20402

UNITED STATES DEPARTMENT OF COMMERCE
Maurice H. Stans, Secretary
NATIONAL BUREAU OF STANDARDS • Lewis M. Branscomb, Director

TECHNICAL NOTE 393

ISSUED OCTOBER 1970

Nat. Bur. Stand. (U.S.), Tech. Note 393, 20 pages (Oct. 1970)
CODEN: NBTNA

A Method for Designing Multi-Screw Waveguide Tuners

M. P. Weidman and E. Campbell

Electromagnetics Division
Institute for Basic Standards
National Bureau of Standards
Boulder, Colorado 80302

NBS Technical Notes are designed to supplement the Bureau's regular publications program. They provide a means for making available scientific data that are of transient or limited interest. Technical Notes may be listed or referred to in the open literature.

For sale by the Superintendent of Documents, U.S. Government Printing Office, Washington, D.C. 20402
(Order by SD Catalog No. C13.46:393), Price 30 cents

CONTENTS

		Page
	Abstract	1
I.	Introduction	1
II.	Tuner Requirements	3
III.	Tuner Analysis	3
IV.	Design Technique	7
V.	References	12
	Appendix A	19
	Appendix B	20

A METHOD FOR DESIGNING MULTI-SCREW

WAVEGUIDE TUNERS

By

M. P. Weidman and E. Campbell

    Capacitive screw, waveguide tuners are commonly used in microwave measurement systems and as devices for adjusting the impedance of various waveguide terminations. The design of a broadband tuner of this type has been a problem in the past.
    This paper describes a method for designing tuners which will work effectively for relatively wide ranges of frequencies.

Key Words: Impedance transformer; waveguide tuner; capacitive screw tuner.

I. INTRODUCTION

The capacitive screw or stub tuner is a familiar device to many who work with rectangular waveguide measurement systems. This tuner consists of one or more cylindrical screws or stubs which can be mechanically inserted into the center of the broad wall of a rectangular waveguide. The stub is then parallel to the electric field for the dominant TE_{10} mode of propagation and appears as a shunting capacitive susceptance for stubs with diameters an order of magnitude less than the wide waveguide dimension [2], [3].

tuner is useful over a full waveguide bandwidth, but for critical measurements it has too much RF leakage. The other form of capacitive screw tuner is the multi-screw type with fixed position screws. The problem in designing the multi-screw tuner is the establishment of a screw spacing such that the tuner is useful over a broad range of frequencies.

The following is a description of a method for determining a spacing of screws in a multi-screw tuner which will be useful at any frequency in a full waveguide bandwidth. To the writers' knowledge, this has never been done before except at NBS. The design technique described here is meant to be an improvement over techniques which have been used at NBS in the past. Tuners have been built in several waveguide bands using general criteria similar to the one described here and have functioned well in many waveguide measurement systems at frequencies throughout the waveguide band. Tuner spacings which do not follow these criteria are, in general, not usable at all frequencies.

II. TUNER REQUIREMENTS

Tuners are used in refined measurement systems to establish idealized impedance conditions. Consider the tuner as a two port waveguide network. If the output port is terminated in a nonreflecting load, and the input port reflection coefficient can be adjusted to any magnitude and phase, then the tuner can be used to achieve any desired impedance condition in a waveguide system.

III. TUNER ANALYSIS

A rigorous mathematical analysis of a single cylindrical stub of variable length extending from the broad dimension of rectangular waveguide is a difficult problem. Mathematical and experimental analyses of this problem can be found in the literature [1], [2].

The analysis used here applies to a multi-screw tuner. Because of the complexity of the multi-screw tuner, several simplifying assumptions will be made:

1) Dominant mode propagation exists in the waveguide everywhere except in the near vicinity of the tuning screws. That is, higher order waves set up by one tuning screw are sufficiently attenuated before reaching another tuning screw.

2) The multi-screw tuner is lossless. This is a good approximation for analytical purposes.

3) Reflection coefficients to be realized at the input port are small in magnitude (less than 0.5).

The model of the cylindrical stub to be used in this analysis is that of a shunting capacitive susceptance. This model was determined experimentally by assuming an equivalent T -- lumped element circuit and then measuring values for the three lumped elements [2]. This equivalent circuit was then converted to an input reflection coefficient (assuming a matched line on either side of the stub) which was then plotted on a normalized Smith admittance chart. The admittance at a single frequency, plotted on a normalized Smith admittance chart, follows the semicircle of unit conductance clockwise as the screw is inserted into the waveguide [3]. This is shown in figure 1. In figures 1-3 it is assumed that only admittances inside the $|\Gamma|$ = .5 circle can be realized. This is done to reduce screw interaction, and more closely approximate the shunt capacitance model for the screw. Tuning also becomes critical as the screw approaches the resonant point (quarter wave insertion) [3].

If the screw is adjusted longitudinally in the waveguide (as can be done with the slide screw tuner) the semicircle of figure 1 is made to rotate through 360 degrees for a longitudinal movement of one-half guide wavelength. From this it can be seen that the slide screw tuner can produce the required range of reflection coefficients.

The multi-screw tuner can be analyzed using the Smith chart along with the three simplifying assumptions mentioned above. A useful multi-screw tuner for a single frequency is one with three screws spaced one-sixth guide wavelength apart. The actual spacings can be one-sixth guide wavelength plus integral multiples of one-half guide wavelength. The range of admittances that can be realized at a single frequency for this three screw tuner is shown in figure 2. In figure 2 the screws are numbered one through three, and the admittances which are realized by adjusting any two screws are those seen at the plane of screw number one or an integral number of half-guide wavelengths from screw number 1. An example of how one admittance would be produced is shown in figure 2 in the region where screws 1 and 2 are used. Screw number 2 is inserted until point B is reached and then screw number 1 is inserted until point C is reached. The reader is referred to Ragan [3] for an explanation of the details of this technique. It can be seen from figure 2 that the requirements for a waveguide tuner, operating at a single frequency, have been met by the three screw tuner with a screw spacing of one-sixth guide wavelength.

Up to this point the analysis has been limited to a single frequency. Figure 3 shows the Smith chart representation of the tuner in figure 2 at a higher frequency. Screws 2 or 3 can be inserted into the waveguide in order to approach admittances in the upper half of the Smith chart, but then screw number 2 will move the admittance away from the area above the screw 1 semicircle. The tuner design shown in figure 2 will be even less useful at still higher frequencies.

The limits on the spacing of a set of screws to be used at any one frequency are then conveniently analyzed with the aid of the Smith chart. In order to be able to converge on an arbitrary admittance by alternate adjustments of two screws, it is desirable to have the locations of the semicircles for individual screw insertions spaced symmetrically around the Smith chart. For good tuner performance, the spacing should be at least as close as that illustrated in figure 2. This would mean that for a broadband tuner there should be no frequency in the band for which there are less than three screws spaced equally around the Smith chart. For actual design this requirement can be altered somewhat.

Another way of specifying tuner spacings is to represent each screw by a phasor. The angular spacings of these phasors would then be

$$\theta_n = \frac{720 d_n}{\lambda_g}$$

where d_n is the distance of the nth screw from some reference point, θ_n is the angular location of the phasor and λ_g is the guide-wavelength at a particular frequency. The design requirement for a broadband tuner is then that at every frequency in the design band there should be no more than 120 degrees between any two adjacent phasors representing points where screws are located. In actual practice this value can be increased to 130 degrees without degrading the performance of the tuner beyond usefulness. It is assumed that the θ_n angles have all been reduced to less than 360 degrees before plotting on a phasor diagram.

IV. DESIGN TECHNIQUE

The design technique described here is only one of many which could be used and still meet the established criterion. Other techniques at the NBS have been used to design tuners which meet the spacing criterion and have been used extensively over the frequency range at which they were designed to operate. These tuners have performed well for various applications. The older techniques did not assure true broadband usefulness; whereas this one does.

In applying the technique suggested here the highest frequency of the design band is considered first. Four stubs are spaced at 120° intervals. Physically this means $\lambda_g/6 + n\lambda_g/2$ at the highest frequency. The $n\lambda_g/2$ or integral number of half wavelengths more than $\lambda_g/6$ spacing depends on the physical characteristics of the tuner. In most applications above 4 GHz $\lambda_g/2$ must be added to $\lambda_g/6$ for spacing since the drive mechanisms used on these stubs cannot physically be spaced any closer together. At even higher frequencies λ_g or more may be needed.

Figure 4(a) shows the stub spacing and figure 4(b) shows a plot of the angular spacing of the first four stubs at the highest frequency. In figures 4(b) - 4(d) the phasor number corresponds to the stub number. Stub number one is always assumed to be at zero degrees (everything is referenced to stub 1).

As the frequency considered becomes lower, the angles between phasor representing stubs are reduced in a clockwise manner as in figure 4(c) until the spacing between phasors 1 and 4 is the maximum 130 degrees. At this frequency, stub 5 is placed so as to be at zero degrees. Stub 5 is placed as close as physically possible to the right of stub 4 in figure 4(a) and is still $n\lambda_g/2$ distance from stub 1. It should be noted here that the smallest possible spacing should always be used since the rate of change of angle with respect to frequency increases with increased spacing.

The preceding procedure is now continued down to a frequency where phasors 2, 3, or 4 may pass through zero degrees into quadrant IV. Up to this point the angles between phasors 2 and 1, phasors 3 and 2, phasors 4 and 3, etc. are all decreasing angles and are less than 130°. At the frequency in which a phasor crosses from quadrant I to quadrant IV, it is necessary to analyze the position of the last stub to be added to the design. If the smaller angle is greater than 130 degrees, another stub must be added at zero degrees. If the angle is less than 130 degrees, the progress of the last stub added is followed down in frequency until the condition in figure 4(d) exists. Note that figure 4(d) is only an example of what can happen. The situation in figure 4(d) is typical of what happens in waveguide bands from WR187 (3.95 - 5.85 GHz) to WR28 (26.5 - 40.0 GHz) with the exception of possibly one less stub involved. Now phasor 8 is moving away from 4 as the frequency decreases. This means another stub (9) must be placed $n\lambda_g/2$ from stub 4. Stub 9 will be located as close as possible to stub 8 and still be an integral number of half-guide wavelengths from 4.

Lower frequencies are considered again until another opening of 130 degrees occurs between phasor 9 and the closest following phasor. It is then necessary to add another stub. Eventually the lowest frequency of the band will be reached. Other phasors will rotate into problem areas thus eliminating the necessity of adding certain stubs. At the lower end of the band, stubs 1 through 4 become sufficient to meet the requirements in the waveguide bands from WR187 to WR62. Above WR62 stubs 1 through 6 play the same role. There will be more stubs than necessary to cover most

frequency ranges using this technique, but if true broadband
usage is desired, all the stubs are necessary.

An example of the design technique for WR90 (8.2 - 12.4 GHz)
is now given. Figure 5(a) shows the first step with the
associated spacing of stubs in inches. It has been assumed
that stubs cannot be placed any closer together than 1/2 inch
due to mechanical considerations. Figure 5(b) shows the
addition of stub 5 to fill the gap at 11.6 GHz. Figures 5(c)
through 5(e) show additional stubs. At 9.87 GHz the conditions of figure 5(f) exist and stub 9 is added. Figure 5(g)
shows the conditions at 9.53 GHz where stub 10 is added.
At approximately 9.21 GHz, phasor 10 is just crossing into
the fourth quadrant at zero degrees. At this frequency
phasor 6 is at 130 degrees and closing the angle with phasor
1. Figure 6(a) shows this condition. No additional stubs
are needed at this point. If the progress of the phasors
are now followed down in frequency it is found that no additional stubs are needed to finish the design. Figures 6(b) -
6(f) show the locations of the phasors for increments down
to 8.2 GHz.

This then completes the WR90 design. There will be no
frequency in the range of frequencies from 8.2 to 12.4 GHz
where an angle of greater than 132 degrees exists between
adjacent phasors representing stub locations. There may be

-10-

the frequency in this design where an angle of 132 degrees between phasors occurs, but the use of an additional stub here is not necessary since it would only mean a small benefit at that one frequency. The tuner will still be usable at that frequency.

A time-shared computer, and BASIC programming language, is used as an aid in the previous design technique. A program (Appendix A) was written which calculates angles for screw spacings and reduces them by integral multiples of 360 degrees to be less than 360 degrees. The output is a print out of these angles, $T(I,J)$. This program needs, as data, the distance of each screw from screw number 1 in inches, $B(I)$, the waveguide width in inches, W, the number of screws, N, and the frequencies (in GHz) at which phasor diagrams are required, $F(I)$. K is the number of frequencies required. A second program (Appendix B) steps through small increments, A, of a desired frequency range, F1 to F2, calculating the difference angles, D, between adjacent screw phasors and lists the value of this angle if it exceeds some predetermined value, C, (for example 130 degrees). The output is a print out of the frequency, F, and the angle, D, if D is greater than C. Frequencies are in GHz. The second program is useful for checking tuner designs or analyzing tuners which have already been built.

V. REFERENCES

[1] Lewin, L. (1951), Advanced Theory of Waveguides, Chapter 4, (Iliffe and Sons, Ltd: London, England).

[2] Marcuvitz, N. (1951), Waveguide Handbook, Chapter 5, (McGraw-Hill Book Co., Inc., New York, N. Y.).

[3] Ragan, G. L. (1948), Microwave Transmission Circuits, Chapter 8, (McGraw-Hill Book Co., Inc., New York, N. Y.).

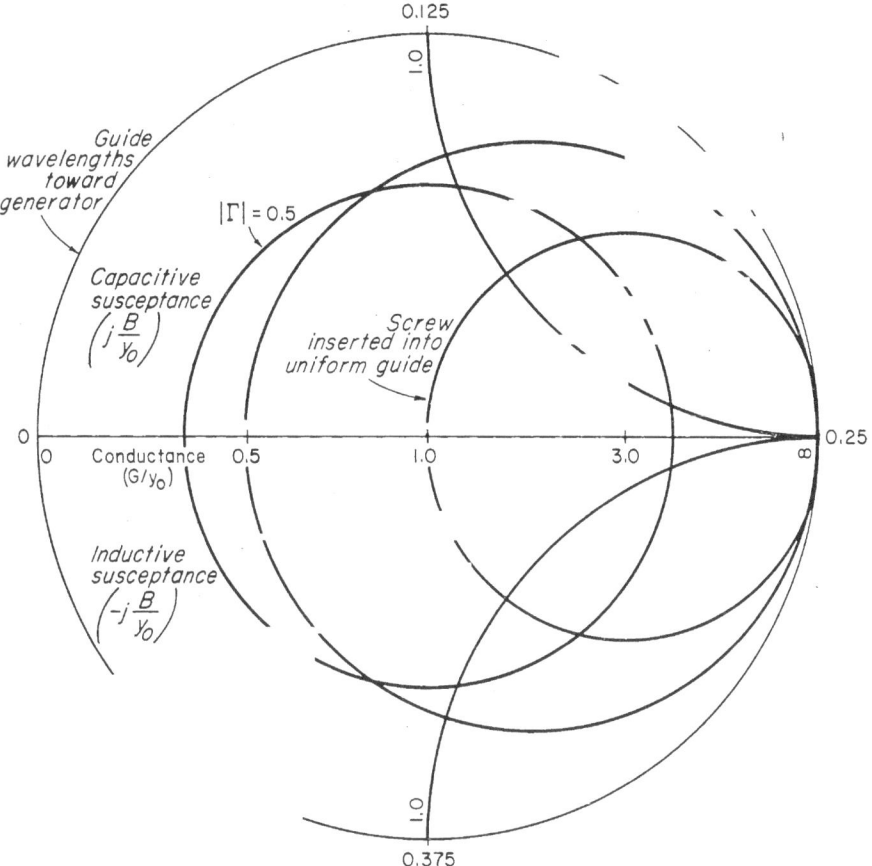

Figure 1. Normalized Smith admittance chart representation of a screw in a uniform waveguide.

-13-

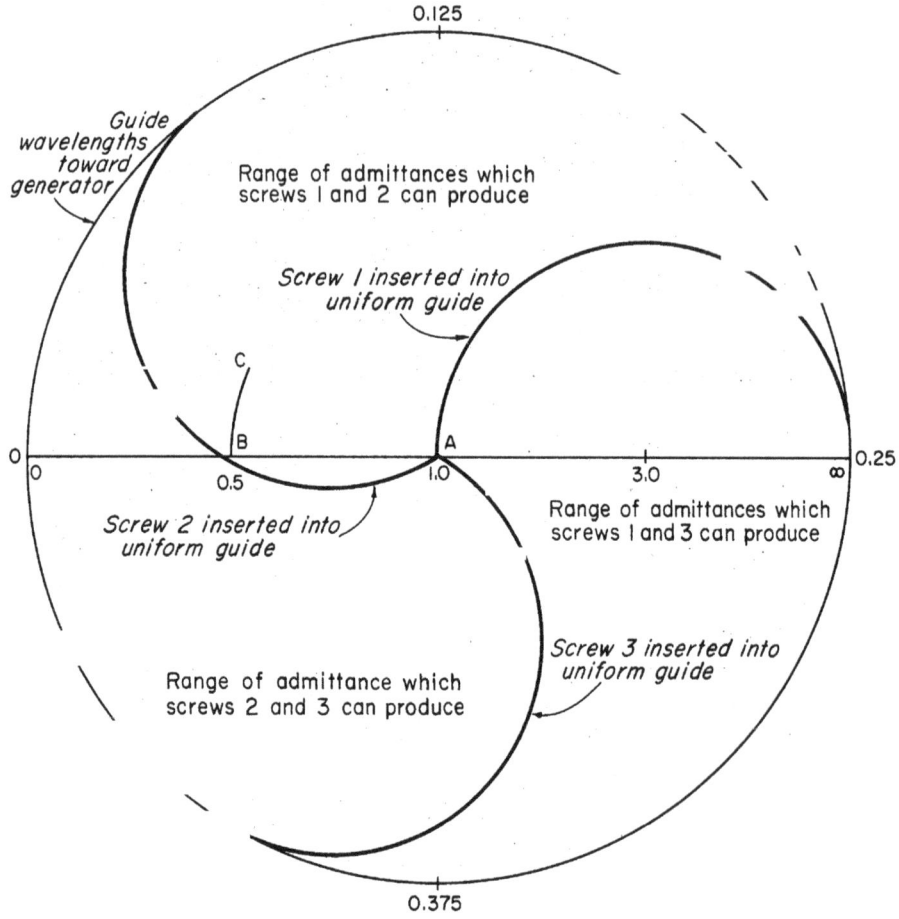

Figure 2. Normalized Smith admittance chart representation of the tuning capabilities of 3 screws spaced one-sixth guide wavelength apart at design frequency and reference plane at 1st screw.

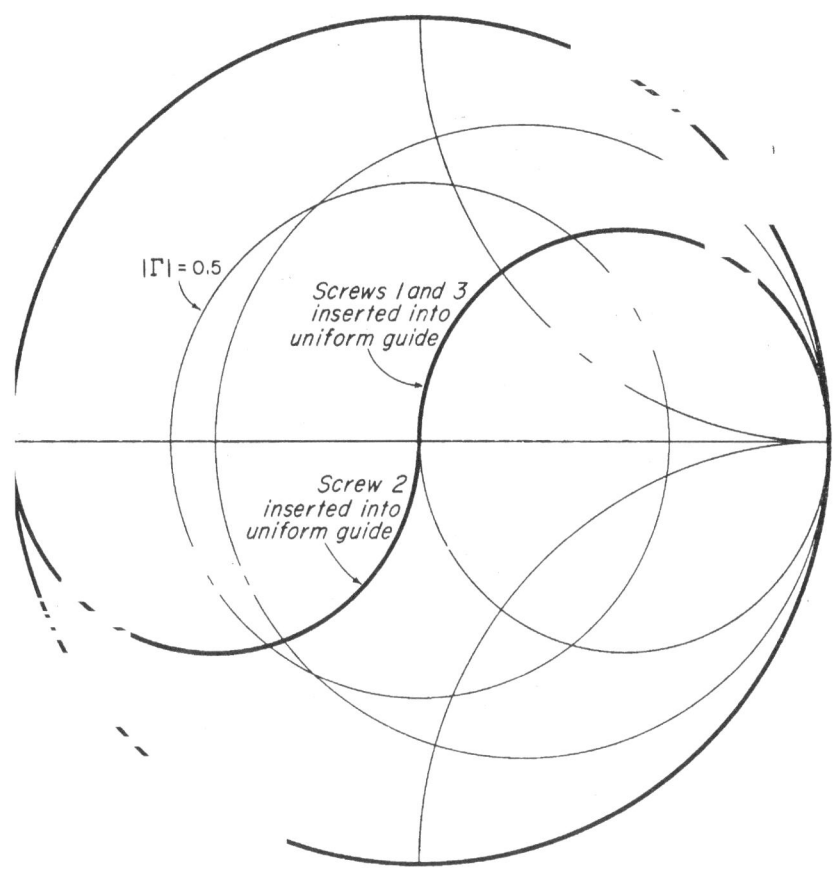

Figure 3. Normalized Smith admittance chart representation of the tuning capabilities of 3 screws spaced one-fourth guide wavelength apart.

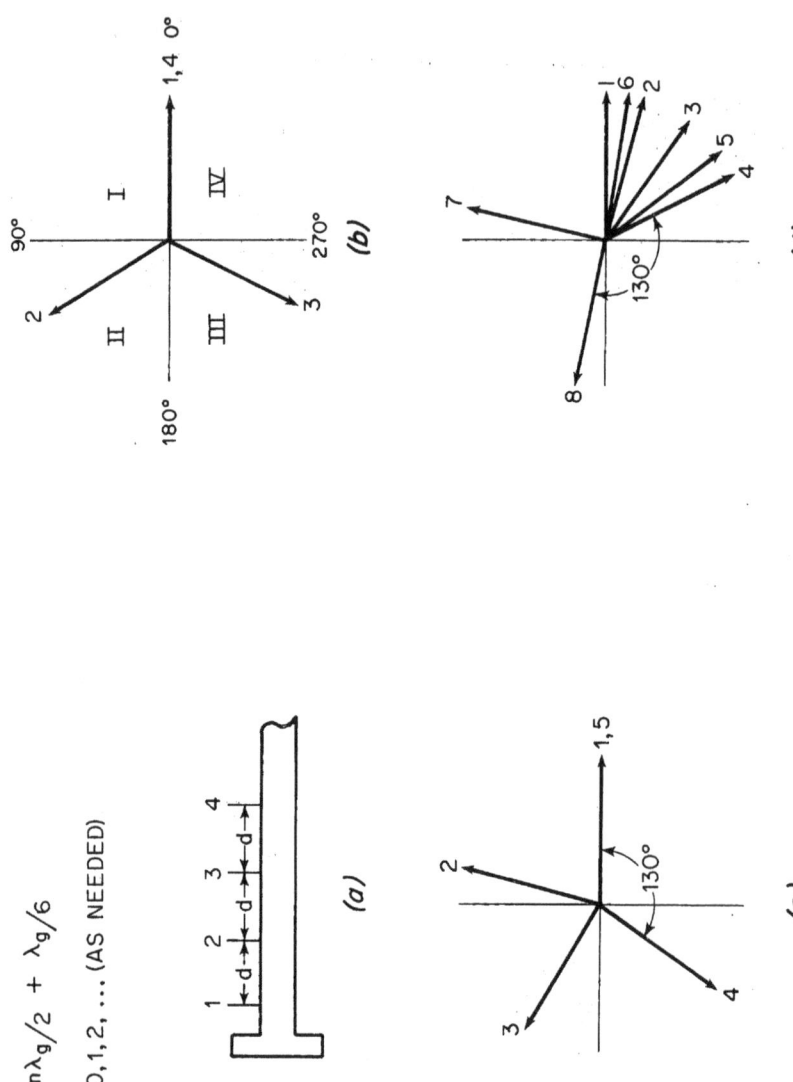

Figure 4. Phasor diagrams.

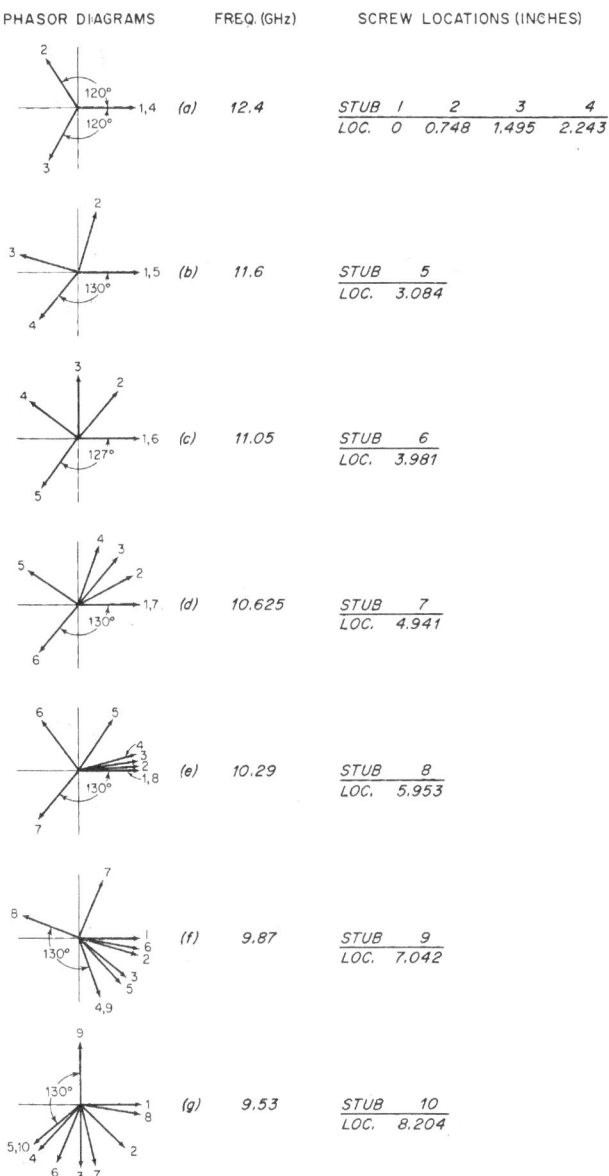

Figure 5. Phasor diagrams.

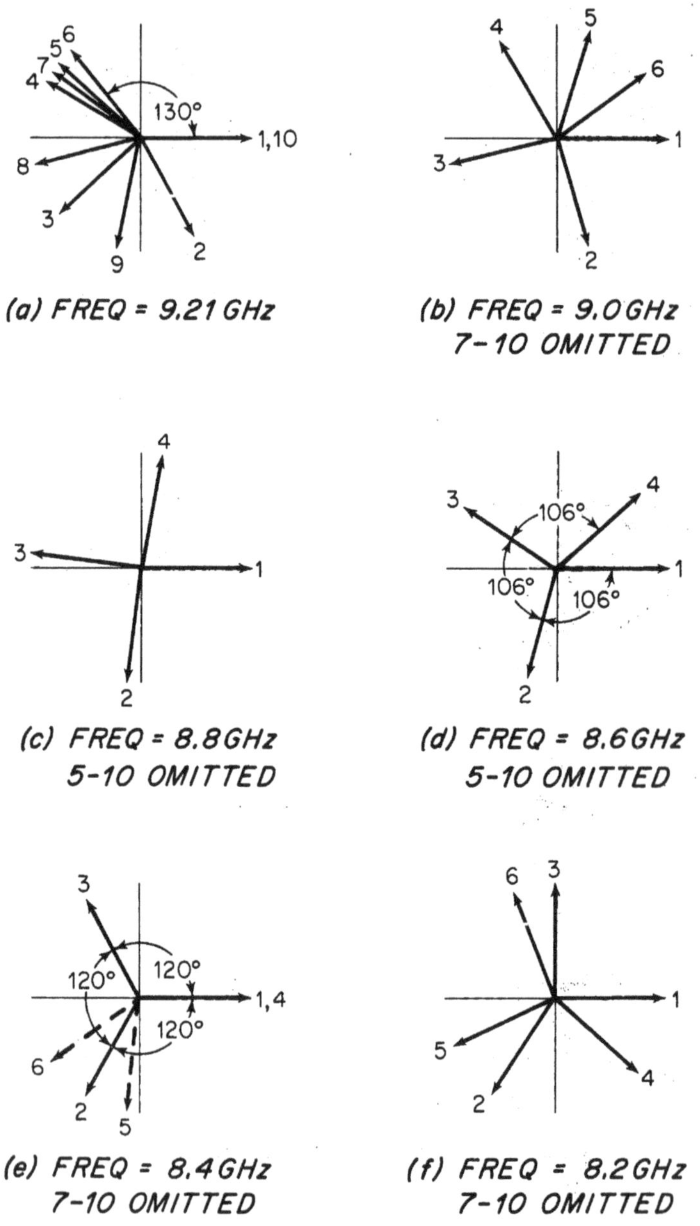

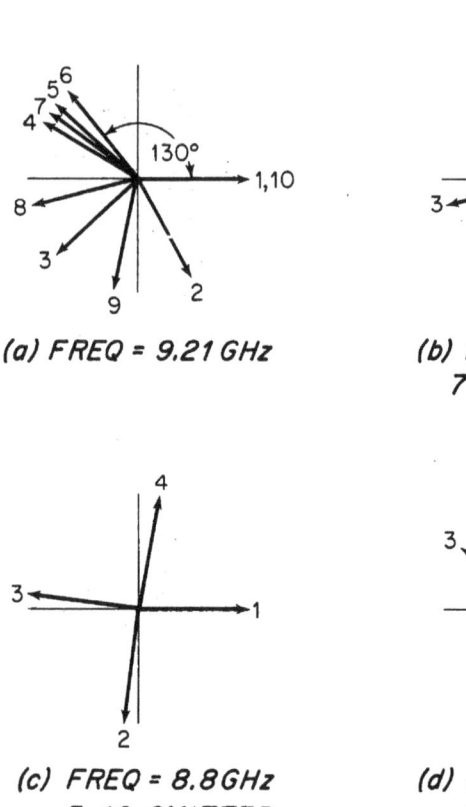

1,10
2
3
4
5
6
7
8
9
130°
(a) FREQ = 9.21 GHz
1
2
3
4
5
6
(b) FREQ = 9.0 GHz
7–10 OMITTED
1
2
3
4
(c) FREQ = 8.8 GHz
5–10 OMITTED
1
2
3
4
106°
106°
106°
(d) FREQ = 8.6 GHz
5–10 OMITTED
1,4
2
3
5
6
120°
120°
120°
(e) FREQ = 8.4 GHz
7–10 OMITTED
1
2
3
4
5
6
(f) FREQ = 8.2 GHz
7–10 OMITTED

Figure 6. Phasor diagrams.

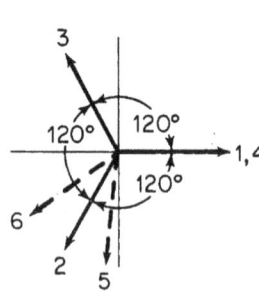

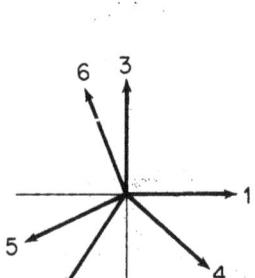

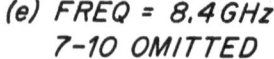

APPENDIX A

```
10   DIM B(15), T(15,20), F(20), G(20)
20   READ N, K, W
30   FOR I = 1 TO N
40   READ B(I)
50   NEXT I
60   FOR J = 1 TO K
70   READ F(J)
75   LET G(J) = 1/SQR((F(J)/11.8027)↑2-1/(4*W↑2))
80   FOR I = 1 to N
90   LET T(I,J) = 720*(B(I)-B(1))/G(J)
100  IF T(I,J) > 360 THEN 130
105  LET T(I,J) = INT(T(I,J)+.5)
110  PRINT T(I,J)
120  GO TO 150
130  LET T(I,J) = T(I,J)-360
140  GO TO 100
150  NEXT I
160  PRINT
170  NEXT J
```

APPENDIX B

```
10   DIM B (15), T (15)
20   READ N, W, C
30   READ F1, F2, A
40   FOR I = 1 TO N
50   READ B(I)
60   NEXT I
70   FOR F = F1 TO F2 STEP A
80   LET G = 1/SQR((F/11.8027)↑2-1(4*W↑2))
90   FOR I = 1 TO N
100  LET T(I) = 720*(B(I)-B(1))/G
101  IF T(I) > = 0 THEN 110
102  IF ABS (T(I)) > 360 THEN 105
103  LET T(I) = 360+T(I)
104  GO TO 110
105  LET T(I) = T(I)+360
106  GO TO 102
110  IF T (I) > 360 THEN 140
120  LET T(I) = INT (T(I)+.5)
130  GO TO 160
140  LET T(I) = T(I)-360
150  GO TO 110
160  NEXT I
170  LET K = N
180  LET S = 360
190  FOR I = 1 TO K
200  IF T(I) > S THEN 230
210  LET S = T(I)
220  LET I1 = I
230  NEXT I
240  LET T(I1) = T(K)
250  LET T(K) = S
260  LET K = K-1
270  IF K > 0 THEN 180
280  FOR I = 1 TO N-1
285  LET J = I+1
290  LET D = T(I)-T(J)
300  IF D < C THEN 320
305  PRINT "FREQ" F
310  PRINT D
320  NEXT I
330  LET D = T(N)+360-T(I)
340  IF D < C THEN 370
345  PRINT "FREQ" F
350  PRINT D
360  PRINT
370  NEXT F
```

Lightning Source UK Ltd.
Milton Keynes UK
UKHW031807150119
335176UK00013BA/1841/P